How To
LIVE ON
MARS

How To

LIVE ON MARS

By CLIVE GIFFORD

Illustrated by
Scoular Anderson

FRANKLIN WATTS
A Division of Scholastic Inc.
New York Toronto London Auckland Sydney
Mexico City New Delhi Hong Kong
Danbury, Connecticut

First published 2000 by Oxford University Press
Great Clarendon Street, Oxford OX2 6DP

First American edition 2001 by Franklin Watts
A Division of Scholastic Inc.
90 Sherman Turnpike
Danbury, CT 06816

Catalog details are available from the Library of Congress
Cataloging-in-Publication Data

ISBN 0-531-14647-2 (lib. bdg.) 0-531-16201-X (pbk.)

Printed in China

Contents

HOW TO LIVE ON MARS

Why go to Mars? The first answer is that it's there. And the second is because—just maybe—we can. We've been up in space. We've landed on the moon. So what's the next step? Mars. And we've already sent unmanned probes to check it out.

But it's been twenty-nine years since anyone went to the moon—why should Mars come up now? What's changed? Two things. One is our technological know-how. Computers today are thousands of times more powerful, and a great deal smarter, than those used on the early *Apollo* missions to the moon. The other big change is what we know about space. In 1998, the *Lunar Prospector* mission saw signs that might lead to the discovery of ice on the moon. Since water can be made into rocket fuel, we might be able to use the moon as a base for manned missions to Mars.

If you're planning a trip to Mars, you might want to ask some important questions:

- ○ Is there life on Mars?
- ○ Was there life there in the past?
- ○ Can we create a human colony on its surface?
- ○ Will there be a space bus to Mars in the near future?
- ○ If there is a space bus, how do I go about booking a seat on it?

This book will answer all these questions and more.

Please fasten your seat belts. Our destination is Mars colony ZXC 0297. Expected flying time, six months.

CHAPTER ONE

WHY LEAVE EARTH?

Human beings are inquisitive. There are always people who want to know what's around the next corner, beyond the horizon, or across the sea. Explorers have traveled to the far reaches of Earth. Mountaineers have climbed the highest mountains, and divers have made it to the deepest parts of the ocean. So what's left to explore? Space, that's what!

After hundreds of years of studying the night sky with telescopes, people became very excited when they were able to send machines, then people, into space.

Yippee!

In the 1950s, scientists put the first artificial satellites in space. Within a few short years, astronauts were sitting aboard cramped spacecraft that orbited the Earth. In 1969, the first person landed on the moon. All this happened very fast, but the next steps—establishing a base on the moon or sending manned missions to Mars—are taking much longer. There are two reasons for this. First, scientists are still working on how to send people as far as Mars. Second, a manned mission to Mars will be very expensive.

Some people believe that sending people and machines into space was, and continues to be, a complete waste of time. We didn't find much on the moon, they say, and it's never going to be possible for lots of people to live comfortably there—or on any of the planets.

Space travel is a very expensive hobby for governments to spend their money on when they have plenty of business on Earth.

But our travels into space have taught us a great deal about the Earth, the universe, the moon, the planets around us, life, and ourselves—knowledge we couldn't have learned any other way.

Spacey Spin-Offs

Critics of space exploration say that all we've achieved from our space missions are thermal blankets and antigravity pens that can write upside down. But the the space program has had many benefits. Research into space has led to direct advances in medicine, robotics, and new materials. Indirect advances have occurred when the needs of the space program drove other industries to improve. The best example of this is the computer industry. When the space program took off in the 1960s, computers were the size of a large room and were very slow. The need for small and powerful computing on spacecraft helped lead to the development of the silicon chip at the end of the 1960s. This is the tiny integrated circuit that makes computers what they are today—small and super-powerful.

The biggest benefit of space exploration has been the invention of satellites, which orbit the Earth thousands of miles above our heads and perform many useful jobs. Weather satellites help us predict weather patterns, storms, and hurricanes. Communications satellites beam live TV pictures and thousands of phone calls all around the world.

Who knows what benefits the new International Space Station will bring once it starts working in a couple of years? Will there be better treatments and cures for certain diseases? Brand new materials? We'll have to wait and see. Curious yet?

Scientists around the world are eager to answer these questions. Curiosity is the main reason why the space station is being built.

Curiosity is also the main reason why there are more and more missions to Mars. Still—why Mars? Well, we've sent machines and people to the moon. Mars is a clear next step. Read on and you'll find out why.

CHAPTER TWO

HI, I'M MARS

The fourth closest planet to the sun, Mars is one of Earth's next-door neighbors (the other is Venus). But sometimes it takes a while to get to know your neighbors.

This is what we've found out about Mars so far—but we'd really like to check in for a closer look.

Some Martian facts:

- About 4,200 miles (6,794 km) in diameter, Mars is about half the size of Earth.

- A day on Mars is just half an hour longer than a day on Earth. But Mars takes

almost twice as long as Earth to orbit the sun. One Mars year equals 687 Earth days.

○ Mars is more like Earth than any other planet in the solar system. This is one reason why many people are very interested in it. Compared to the other planets in our solar system, Mars is remarkably friendly.

PLANETARY FACT FILE

Mercury: Forget it—far too close to the sun!

Venus: Heavily clouded, but still so hot that it would melt lead. Also, the atmosphere is about ninety times as thick as Earth's and is mostly made of harmful carbon dioxide.

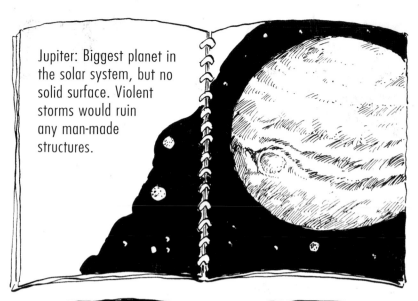

Jupiter: Biggest planet in the solar system, but no solid surface. Violent storms would ruin any man-made structures.

Saturn, Uranus, and Neptune: these planets lack a solid surface to land on. Also, they are very far away, and are surrounded by clouds.

Pluto: Too far away.

Compared to icy-cold Neptune and Uranus, boiling-hot Mercury, and just plain scary Venus, Mars looks like a vacation spot. But let's take a closer look below the surface and see what we find.

Desert—Rats! ....

Early astronomers were very excited when they looked at Mars's surface through telescopes and saw large red-orange areas. They thought these were sandy deserts—and where there's a desert, there's bound to be an oasis with trees and water, right?

Wrong!

So far, no telescope, however powerful, has found any trace of a palm tree. The real news is that the surface of Mars looks like desert, but it's ice-cold. The average temperature on Mars is -73.4°F (-23°C).

That single temperature doesn't tell the whole story. Mars's temperature varies greatly throughout the year.

At its warmest, the area around Mars's equator can reach a sultry 80.6°F (27°C), while the lowest recorded temperature on the planet is -262.4°F (-128°C). In a single day, Mars can have temperature differences of up to 212°F (100°C).

Mars's surface does look a bit like a desert, though. Loose stones, soil, and sand cover much of it. The soil on Mars is made up mainly of silicon and oxygen, mixed with metals such as magnesium and iron. When iron and oxygen react together, they form iron oxide, or rust. This is what gives Mars its distinctive red color.

Move to Mars
MAKE YOUR OWN MARTIAN SOIL

WHAT YOU'LL NEED
* steel wool
* a saucer
* rubber gloves
* a paper towel

WHAT TO DO
1. Fold the paper towel in half twice and place it in the saucer.

2. Run warm water over the steel wool for a while. Then place the wet steel pad on the paper towel, and put the saucer in a place that won't be disturbed for a week or so.

3. After two weeks, put on the rubber gloves and rub the pad between your fingers.

WHAT HAPPENS?
The pad should disintegrate into a reddish powder. This is iron oxide, or rust.

Atmosphere.................................

The atmosphere of Mars is made up of carbon dioxide, nitrogen, argon, oxygen, water vapor, and carbon monoxide. Because water is vital to life, scientists have become very interested in precisely how much water there is in the Martian atmosphere. The answer is not very much. If you could take the entire atmosphere of Mars and wring it out, the water would fill a few dozen swimming pools. Recent evidence shows that Mars might have two to three times this much water, however.

A few dozen swimming pools may sound like a lot, but compared with the amount of water in the Earth's atmosphere, it's almost nothing.

Missions to Mars ....

There have already been well over twenty Mars missions. Unfortunately, many have been complete disasters. The former Soviet Union launched eight missions to Mars in the 1960s, some "landers" and others fly-by missions. Not one of them made it.

Oh no, missed again!

The first data about the atmosphere and surface of Mars came with the *Mariner 4* spacecraft, launched in 1964. This was the first successful fly-by of Mars. *Mariner 4* traveled to within about 6,200 miles (10,000 km) of the planet's surface, found that carbon dioxide was the main gas in the atmosphere, and took 22 photographs. It took *Mariner 4* eight hours to send each fuzzy picture back to Earth. Still, scientists were very excited. They had in their hands the first close-up pictures of Mars's surface.

Whipping Up a Storm ....

With no rain or surface water, Mars's surface is as dry as a bone. Storms on Mars often whip up the soil and sand and create enormous red dust clouds. These can cover large parts of the planet—sometimes even all of it—for months at a time. Early astronomers saw dark patches on Mars and hoped these were lush forests. But sadly, they weren't. The darker patches are areas where the winds have blown away the loose material on the surface to reveal the solid rock beneath.

Move to Mars
SEE HOW WINDS SHIFT DRY MARTIAN SOIL

You can see in the following experiment how winds move around the loose, dry material on the surface of Mars.

WHAT YOU'LL NEED
* a pan or deep tray
* a hole puncher
* a piece of paper
* a plastic spray bottle

WHAT TO DO
Punch about fifty holes in the paper and empty the small paper circles from the hole puncher into the pan. Blow across the paper circles. Now spray the paper circles with water from the spray bottle until they are damp. Blow across the circles again.

WHAT HAPPENS?
The dry paper circles move easily when you blow them. When they are damp the circles are heavier, and they stick together, so they are harder to move.

Caps Off

Mars has two polar ice caps that seem to grow in the winter and shrink in the summer. The ice cap at the north pole is bigger than the one at the south pole.

620 miles (1,000 km) wide

186 miles (300 km) wide

This ice is not necessarily frozen water. Scientists hope it is, because if so, future missions might be able to melt part of the caps to provide water. However, until a mission successfully examines the area and lets us know, we can't be sure. Many scientists think the caps are made at least partly of frozen carbon dioxide. On Earth we can see a form of frozen carbon dioxide in action. It's called dry ice, and it often provides the "smoke" in movies and at rock concerts.

THE MONSTER FROM MARS

Valley Deep, Mountain High..................

A number of unmanned missions to Mars have allowed us to map the planet's surface. Two of Mars's most prominent features are a giant inactive volcano and a huge valley. The Valles Marineris (Mariner Valley) makes the Grand Canyon look like a hole in the road. It's a staggering 372 miles (600 km) wide and 2,800 miles (4,500 km) long. That's as long as the distance across the United States.

Olympus Mons, Mars's highest mountain, rises about 14 miles (23 km) above the ground. That's just short of three times the height of Mount Everest. The crater at the top of Olympus Mons could easily swallow New York City.

Olympus Mons and the Valles Marineris were
photographed for the first time by the *Mariner 9*
spacecraft. *Mariner 8* had been a
complete failure—it plopped
into the sea just off the island
of Puerto Rico. But *Mariner
9* was a great success.
It took 7,329
photographs of
parts of Mars's
surface.

No Pressure ....

The atmosphere on Earth is very heavy. Every cubic yard of air weighs about 2 pounds. That's the same as having 2 pounds pressing down on every inch of your body all the time!

Move to Mars
SEE ATMOSPHERIC PRESSURE IN ACTION

WHAT YOU'LL NEED
* ✳ a flexible plastic bottle
* ✳ a deep breath

WHAT TO DO
Breathe out completely, then put your lips firmly around the bottle opening and breathe in deeply. You're sucking the air out of the bottle.

WHAT HAPPENS?

The bottle buckles as its sides go in. What is forcing the sides in? Air pressure! You're sucking air and air pressure out of the bottle, causing the air pressure outside the bottle to become greater than that inside the bottle. That's what forces the sides in.

The atmosphere on Mars is incredibly thin. It exerts much less pressure on the planet's surface—less than one-hundredth of the pressure here on Earth.

A Matter of Some Gravity......................

Mars has some gravity, but not as much as Earth. The gravity on Mars is about one-third of what we experience on Earth. This would make astronauts great at golf because golf balls would travel much farther. More important, lightweight, inflatable buildings and structures could be built easily.

But we're getting ahead of ourselves. Before we land on Mars and start building, let's look at how other people have seen Mars throughout history.

CHAPTER THREE

THE LURE OF THE RED PLANET

Compared with the other objects in the sky, Mars has always stood out as a bit unusual. It glows red, while most of the stars are white. It also makes a strange looping movement across the night sky. The ancient Egyptians called Mars *seked-ef em khetkhet*—"one who travels backward." Mars does appear to travel backward for a short time each year, but today we know that isn't really what happens. Because Earth is closer to the sun than Mars, its orbit isn't as long. So, when the Earth "overtakes" Mars as it orbits the sun, it looks as if Mars is going backward.

Fortunately for us, our glowing red star later got a much shorter name—one that's much easier to spell. The red planet was named after the god of war—Ares to the Ancient Greeks, but Mars to the Romans. Many Romans worshiped the god of Mars as their protector. Some thought the planet's red color came from running blood—scary!

New Ideas . . . .

Nothing much more happened with investigating Mars until the sixteenth and seventeenth centuries. By this time, the printing press had been invented, and many more people were reading books. This triggered off an explosion in thought and learning, much of it about the stars and planets in the sky. People read the works of the ancient Greeks and Romans and discussed new ideas. Unfortunately, if their ideas angered the churches or rulers where they lived, they were in serious trouble.

For instance, at that time most people thought the Earth was at the center of the universe, and that the sun, moon, stars, planets, and everything else circled around it. A few people suggested that maybe we'd gotten it wrong. Maybe the sun was the center, and the Earth, Mars, and all the other planets circled around it. They were right, of course—but hardly anyone wanted to know, especially not the Church. And in those days, the Church could take all sorts of measures to keep people from talking.

The Dueling Dane

Denmark, 1576

The first great Mars explorer was a Danish nobleman named Tycho Brahe. He was completely devoted to astronomy but still found enough time to lead a colorful life. He lost much of his nose in a dueling accident and fitted himself with a new one he made out of gold, silver, and wax.

Who are you calling a fool to study the stars?

Old Metal Nose was a remarkable astronomer. In 1576, he set up an observatory on an island near Copenhagen. For the next twenty years, he observed and kept detailed records of more than seven hundred stars. His calculations of the positions of planets, especially Mars, were incredibly accurate. All the more amazing was that Brahe did all this before the invention of the telescope. What did he use, then? His eyes, that's all.

Brahe had a student named Johannes Kepler. They didn't get along very well. As a result, Brahe gave Kepler some really tough homework: studying the movement of the solar system's most erratic planet—Mars.

It's the movement of Mars for you!

Oh no! I'd prefer more math to that!

After Brahe died in 1601, Kepler discovered that Mars travels around the sun in an oval-shaped orbit called an ellipse. He also calculated that Mars might have two moons, but never got to see them. Poor Kepler wrote to the Mr. Telescope of his time, Italian scientist Galileo Galilei. He begged and pleaded for a loan of one of Galileo's "seeing machines," but the Italian refused.

Handy Hans
Holland, 1608

"Yikes, a giant rat!"

The telescope had been invented in 1608 by a Dutch eyeglass-maker named Hans Lippershey. Fixing lenses in a row inside a tube, he found that he could make things appear several times bigger than they really were.

Great Galileo

Italy, 1609

It was Galileo who really developed the telescope. Just a year after Lippershey's invention, Galileo built a telescope that could magnify thirty times. With the help of his telescopes, Galileo made many important discoveries about our solar system.

What I discovered today:
1 There are four moons around Jupiter.
2 The Earth's moon is not smooth but full of mountains and craters.
3 The Milky Way is made up of lots of individual stars.
4 Venus has phases like the Moon.

Telescopes gradually got bigger and better. Much thanks is due to a Dutch scientist, Christiaan Huygens. He found ways of grinding and polishing lenses to make them work better. He used his new, improved telescope to discover a moon around Saturn. Huygens was the first person to produce a sketch of Mars that had correct features on it.

Moons of Mars ....

Remember that prediction of Kepler's? You know, the one where he thought there were two moons orbiting Mars? Well, it took more than two hundred years to prove him right. In 1877, an American astronomer named Asaph Hall discovered both moons. Given the honor of naming the moons, Hall studied Greek myths. The Greek god of war (from whom Mars got its name) was always joined in battle by his two children, Deimos and Phobos. That's what Hall named the two moons.

Canal Fever ....

1877 was a big year for Mars watchers. Aside from the discovery of the planet's moons, there was also an announcement from Italian astronomer Giovanni Virginio Schiaparelli. He noticed straight lines cut into the surface of Mars, which was correct. But what followed was a bit of a mistake.

Words often have many meanings, so they can get translated incorrectly. In Italian, Schiaparelli had written "channels," but in English it was translated as "canals." The Suez canal was being built at the time, and people were canal-crazy. To them, canals meant creature-made waterways—which meant there had to be life on Mars!

We now know the "canals" that caused so much fuss were imaginary, or perhaps optical illusions. There are channels cut into the Martian surface, but when astronomer Patrick Moore compared Schiaparelli's canal maps with the maps of Mars we have today, they didn't fit at all. Scientists think these channels were probably formed by water and ice eroding the rock. This means Mars was once a planet on which water flowed.

Move to Mars
SEE HOW ICE HAS FORCE

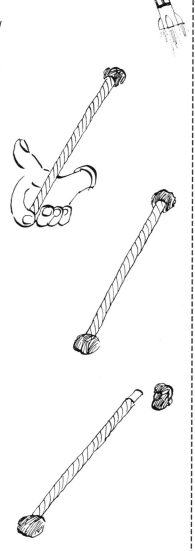

WHAT YOU'LL NEED
* a plastic drinking straw
* some modeling clay
* a glass of water

WHAT TO DO
1. Suck some water up into the straw. Put your finger over the end of the straw, so that the water stays in while you plug the open end with some of the modeling clay.
2. Turn the straw upside down and plug the other end with the rest of the modeling clay.
3. Put the straw in the freezer and wait for three hours. Open the freezer and look at the straw.

WHAT HAPPENS?
The water has turned to ice, and one of the clay plugs has been forced out. When water turns to ice, it expands and pushes with force. When water gets into cracks in rock and freezes, it can push hard enough to crack the rock.

Astronomers, scientists, and the general public couldn't stop talking about civilizations on Mars. Martians became a popular theme for books, comics, cartoons, and films. H.G. Wells's book *War of the Worlds*, which tells of Martians landing on Earth and taking over, was a best-seller. Later, a retelling of the story became the most famous hoax of all time.

Invasion from Mars!

The United States, October 30, 1938

It didn't happen on April Fool's Day, but on the eve of Halloween in 1938. Orson Welles staged a radio broadcast based on *War of the Worlds*. It was announced as a fictional story, but many people still believed Martians had really landed. Thousands of Americans were panic-stricken. Many fortified their houses and loaded their shotguns to wait for the invasion. It took a few days to convince everyone that it was just a story.

The Martians have landed!

Welles's broadcast turned him into an overnight star. It also showed how many people really believed there might be life on Mars.

CHAPTER FOUR

ALIEN LIFE

People all over the world are fascinated by the idea that alien life might exist. That's why TV programs, movies, and books about aliens are such a success. It's also why books on unidentified flying objects (UFOs) sell like hotcakes.

Did They Land? ....

The first attempt to get a machine onto Mars to find out whether aliens live there happened in 1969. The former Soviet Union had already sent seven missions to Mars. Scientists believe some of these actually made it to the surface, but contact was lost with all seven of them. Perhaps, in the future, we'll find out what really happened.

The HoaX Files ....

Most people who claim aliens have visited Earth in the past have been unmasked as hoaxers. And while we're spoiling the party, we might as well go further. The vast majority of UFO sightings are actually sightings of airplanes, weather balloons, optical illusions, and natural phenomena such as St. Elmo's fire, a type of electrical charge that creates a glow around tall objects. The planet Venus is always being reported as a UFO. People want to believe life from other planets and stars exists, however, so they try really hard to prove it—or fake it.

Move to Mars
FAKE YOUR OWN UFO PICTURE

Let's see how easy it is to make your own fake UFO photograph.

WHAT YOU'LL NEED
- ✳ a camera with film
- ✳ a friend
- ✳ a round metal baking dish

WHAT TO DO
You and your friend have to be outdoors in plenty of space. ➤

Ask your friend to have the camera ready. It should be focused on the background. Gently toss the baking dish as you would a frisbee. Try to throw it at an angle so that the bottom of the dish partly faces the camera—this will give it a more convincing shape in the photo. Take several photos, then take the film to be processed.

WHAT HAPPENS?
When the pictures come back, one, maybe more, should be half-convincing. The object in the sky will be blurred, giving the impression that it is moving fast.

Party Poopers

Scientists aren't heartless monsters with no emotions or passions. On the contrary, many of them went into science because they hoped they could prove something that other people just laughed at. But scientists also have a duty to prove things scientifically whenever they can. Even though most sightings and theories of alien life have proved to be nonsense, there do remain a few unsolved mysteries...

How come ancient people built really clever star observatories and calendars? Surely they had alien help.

What about the many ancient paintings of creatures with what look like space suits on?

What about the Nazca lines in Peru, which look like runways from the sky?

As we've said, no one truly knows, and perhaps we never will. Most people think the lines in the sand and the space suits just came from the imaginations of ancient artists. And there's a good chance that the carvings showing the movements of the sun and stars were the work of dedicated astronomers from ancient cultures. We have to admit that these explanations are more likely than an alien race traveling hundreds of millions of miles and still needing twenty-first-century airport runways to land.

What's that, Capac?

I don't know, just a doodle. It's going to drive future archaeologists crazy!

The Search for Alien Life ....

Most serious scientific searches for alien life look past our solar system and into deep space. There are billions of stars in the universe. Many scientists figure that the chances of life somewhere else in the universe are small but possible—and possible might be just good enough.

One in a billion—that's good enough for me.

The Search for Extraterrestrial Intelligence (SETI) is an organization dedicated to finding life elsewhere. Among its many projects, SETI uses radio telescopes and computers to monitor radio frequencies for any possible radio signals sent to Earth. So far they've had no luck, but they remain hopeful.

This is Radio Andromeda. It's five pulses past sol and here's the traffic check. There's a jam up near Alpha Centauri...

In 1974, SETI sent a radio signal to a cluster of stars 24,000 light-years away. The signal was sent in the form of pulses that made up a picture. The picture contained a number of things about Earth, such as its position in the solar system and what atomic elements make up life on our planet.

Move to Mars
COMMUNICATE WITH ALIENS

Make your own message to send to aliens on another planet.

WHAT YOU'LL NEED
* ✳ a pen
* ✳ a piece of notepaper
* ✳ your brain in high gear

WHAT TO DO
What would you include on a piece of paper to tell a creature who doesn't speak English all about you and the planet you come from? Write it down and share it with a friend.

Pioneer and Voyager

Several space probes have been sent beyond Pluto and into deep space. *Pioneer 10* was the first man-made object to leave the solar system—a journey that took nine years. The ship carried a plaque with pictures of a man and woman, a diagram of our solar system, and other information. By the time the *Voyager* space probe was launched in 1977, video had been invented. The probe carried a videotape that contained photos of Earth from space and various people of our world. It also included sounds, from an elephant's trumpeting to greetings in different languages and music—even the sound of a kiss!

The chances of *Voyager's* videotape falling into the hands of an intelligent life form are very, very slim. It is even less likely that an alien would know what the tape is, play it, and understand it.

It's easy to make fun of alien-watchers, but scientists continue to hope that contact will be made. It would certainly be the biggest event of the millennium.

But What About Mars? ....

From the tantalizing glimpse of Mars that scientists have had from space probes and telescopes, one major fact has emerged. It is 99.9 percent certain that no intelligent alien life exists on Mars today.

How come?

The *Viking* missions that successfully landed on Mars proved that Mars has

○ no liquid water
○ almost no oxygen

○ intense cold
○ no vegetation

Viking 1 and 2

In 1976, two *Viking* spacecraft parachuted safely down onto the surface of Mars. They made continuous measurements of temperature, wind speed, and the length of the Martian day. Robot arms fitted to each lander scooped up soil samples and analyzed them inside the craft. Much of our knowledge about the surface of Mars comes from these two spacecraft. They worked thousands of miles apart until they were switched off in the 1980s.

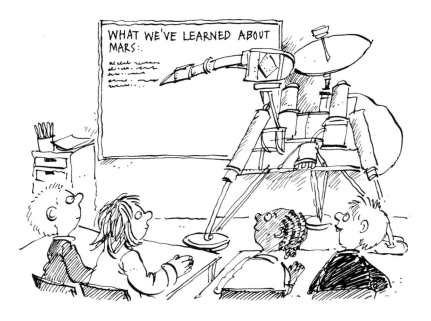

WHAT WE'VE LEARNED ABOUT MARS:.

What keeps the scientific pulse racing about Mars is the question of whether life of any kind existed there in the past. Many scientists believe Mars had lots of water flowing on its surface billions of years ago. The trick now is to find proof of life buried in the Martian rock.

The Martian Meteorite Mystery ....

One rock, a meteorite, caused an incredible stir in 1984. Found in Antarctica, the rock had come all the way from Mars about thirteen thousand years ago.

I travel to Earth to escape the cold of deep space, and look where I end up!

Labeled ALH84001, this chunk of rock appeared in newspaper headlines for weeks. The evidence of bacteria inside the rock appeared to be proof of ancient life on Mars. But not all the scientists agreed. There is a chance that the features inside the meteorite were formed by a geological process, without the presence of bacteria. It is even more likely that they were formed by Earth bacteria after the meteorite landed in Antarctica. This debate is still raging.

Many scientists now believe we need a mission to Mars that can actually return to Earth, in order to bring back fresh rock and soil samples for scientists to analyze. NASA is working on just such a mission, to be launched in about five years.

By the time the mission returns, you might just have finished your astronaut training.

CHAPTER FIVE

GETTING THERE

The universe is a vast place. Most of it is out of our reach at the moment. Let's look at a few examples. If you turned your family car into a star car and traveled at a constant 99 mph (160 km/h), it would take you 221 billion years to reach the center of our galaxy, the Milky Way. It would take you 3,154 years just to get to Pluto, the most distant planet in our solar system. But getting to Mars would be a breeze if you packed a big enough picnic basket—just 39 years and you'd be there. And spacecraft these days can travel a bit faster than your souped-up family truckster.

Sending people to investigate Mars would be a major achievement, and it won't happen without a lot of hard work. Many unmanned missions have failed in the past. And although we've learned from them, and now have far more powerful computers, unforeseen errors can still occur. Just ask the people involved with NASA's *Climate Orbiter*.

Here We Go Again

Mars, September 1999

The *Climate Orbiter* was designed to be Mars's first weather satellite. All was going well until September 1999. The craft had reached Mars's orbit, but something went wrong, and the signals went dead. The spacecraft got too close to Mars and burned up in the Martian atmosphere. One reason for the crash was that one whole team of engineers was using metric units of measurement, while another was using feet and inches!

One advantage of manned missions is the cargo they carry—people. The first Mars voyagers will be supreme technical astronauts able to perform many different tasks, including repairs to parts of the spacecraft that jam, buckle, or simply stop working.

Human cargo is also one of the biggest headaches, however. Unlike robots and automatic experiments, humans need to be fed and watered, supplied with air and toilets, and given somewhere to sleep. This all takes lots of space and weight.

The problem is that it takes a long time to get to Mars. The shortest journey to Mars would probably take about six months. That means carrying at least six months' supplies of food, water, air, and other necessities to keep a spaceship and its crew going...

And all this cargo has to be blasted clear off Earth.

Escaping Earth.................................

It's no easy task to produce enough power to get away from Earth.

You have to reach a speed known as escape velocity to tear the spacecraft free of Earth's gravity, and it's pretty fast—nearly 25,000 mph (40,000 km/h).

I don't think he's going to hit escape velocity.

Scientists have suggested several ideas for launching future Mars missions. One involves launching the mission from the moon, which has only a tiny fraction of the gravity of Earth. This idea got a boost in 1998, when the probe *Lunar Prospector* found telltale signs of ice containing water at the moon's poles. Water is made up of hydrogen and oxygen, and hydrogen and oxygen are just the ingredients for making rocket fuel.

Another idea involves using the International Space Station, being built right now, as a sort of giant construction and launch site. The many parts of the Mars mission craft would be hauled up to the station, assembled there, and then launched from space.

It's far more likely that the first manned Mars missions will use tried and trusted multistage rockets instead. These fire stage by stage and fall away when their fuel is used up, lightening the load to be sent into space. Here's one in action:

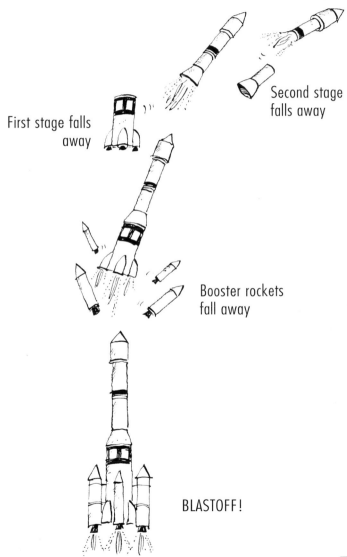

First stage falls away

Second stage falls away

Booster rockets fall away

BLASTOFF!

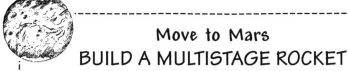

Move to Mars
BUILD A MULTISTAGE ROCKET

Here's how to make your very own multistage rocket, where parts fall away as they use up their fuel.

WHAT YOU'LL NEED
* a paper cup
* a long balloon
* a pair of scissors
* a round balloon

WHAT TO DO

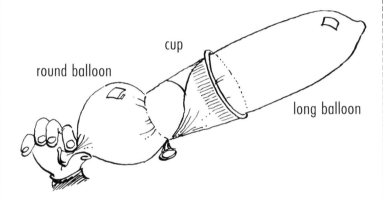

round balloon

cup

long balloon

Cut the bottom out of the paper cup. Blow up the long balloon and pull the open end of the balloon through the top and out the bottom of the cup. Fold the open neck of this balloon over the edge of the cup. This temporarily keeps the air from leaving the long balloon. Place the round balloon inside the cup and inflate it. Let go of the neck of the round balloon.

WHAT HAPPENS?
The round balloon propels your balloon rocket away. As soon as it deflates, the round balloon and the paper cup fall away, leaving the long balloon to speed forward. This set of balloons is just like a multistage rocket.

Weighty Matters . . . .

The exact design of the Mars mission rockets can be left to rocket scientists. What is important is the size and weight of what will be sent into space. Experts call this the payload.

You can forget your hardcover manuals and encyclopedias—especially when ten heavy volumes can fit easily on a CD-ROM. And all those cameras your friends gave you to take pictures for them? Forget it! Unfortunately, you're going to leave your pet parakeets at home, too.

To reduce weight, a manned mission is likely to involve two, three, or even more spacecraft. The craft with the crew on board will be launched on the quickest path available. Before that, robots will play a big part. They'll be sent ahead, traveling second-class on slower cargo spacecraft to Mars. Upon arrival, the robots will set up machinery for making rocket fuel and other supplies. They'll check to make sure all the cargo is fine and radio the message back to base.

Several cargo missions, if successful, would provide all the supplies and equipment that the human crew would need for their time on Mars.

Launch Windows ....

Another important thing about going to Mars: you have to pick the right time to go.

How far Mars is away from us varies. Earth and Mars make their trips on different paths around the sun. There are key times called launch windows when Mars is at its closest to Earth and other conditions are right. These are the best times to start a space mission to Mars. Every 25 months or so, a reasonable launch window occurs. But the very best launch window—when Mars is closest to both the Earth and the sun—comes around every 15 to 17 years. One of these is windows is happening in 2001.

Stay in Touch . . . .

Once free of Earth, the manned craft would begin its
five- or six-month trip on a carefully planned flight
path to Mars. As it travels farther away from Earth, it
will take longer to send and receive messages. On
Mars, the delay reaches ten minutes, as the successful
1997 *Pathfinder/Sojourner* mission to Mars proved.

Pathfinder and Sojourner

Mars, July 1997

Once it landed safely on Mars, the *Pathfinder* probe opened like
a flower. Its petals were solar panels that had also been
Sojourner's garage on the seven-month trip to Mars. *Sojourner*
was the little rover that wandered around a tiny part of the
Martian surface, investigating rocks and soil.

➤

Sojourner sent information back to *Pathfinder*, and *Pathfinder* forwarded it to Earth by means of radio waves.

Landing Party

Upon reaching Mars, the craft would go into orbit around the planet before timing its descent and landing. Cushioned by parachutes, rocket blasters, and possibly even large balloons, the craft would land at its target area. A successful landing would cause huge celebrations back at mission control.

CHAPTER SIX

LIVING ON MARS

Mars would be a great place to live if it weren't for the lack of air, food, and plants—not to mention the intense cold. How would it be possible for people to get by on Mars, even for a short time? And, more important, could anyone actually live there? It will take a lot of ingenuity, but experts think it is possible with the technology we already have. Information from unmanned missions such as *Global Surveyor* will be vital in planning how people can survive on the red planet.

Global Surveyor

In orbit around Mars, right now

Launched in 1996, *Mars Global Surveyor (MGS)* was overshadowed by the *Pathfinder* mission. This is a shame because, despite some minor problems, *MGS* has produced plenty of information about Mars—data that could be very important to all the early visitors. Packed with high-tech equipment such as radar, lasers, and advanced cameras, *MGS* was the first spacecraft to analyze Mars's weather, how the polar ice caps grow and shrink, and lots of other things. It will certainly keep plenty of scientists busy.

Don't tell me: more data from <u>Mars Global Surveyor</u>.

Build Your Own Extension ....

You and the other astronauts on a short mission to Mars should be able to get by living in your landing craft. This will probably be fitted with a couple of small, pressurized extensions that look a bit like high-tech trailer awnings.

The key to these extensions and to larger buildings on Mars will be to keep them full of air under pressure. This means astronauts can live inside them without cumbersome pressure suits. Astronauts will enter and exit through airlocks that prevent air from escaping and keep the inside from becoming contaminated with dust.

Suits You

To go outside on Mars, you will have to wear an advanced space suit built for the unique Martian conditions. "Marswear" will have to be a lot lighter than the types of space suits and backpacks worn by astronauts visiting the moon. Mars's gravity may be less than Earth's, but it's double that of the moon's.

Engineers are hard at work developing a lightweight suit that can be packed with enough supplies for four or more hours in the Martian atmosphere. Food, water, power, communications, and a breathing device all have to be crammed in. The suit must also allow the astronaut to move freely. It's a tall order, but essential to a Mars mission.

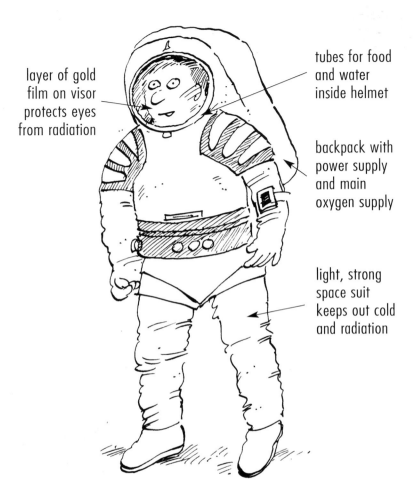

layer of gold film on visor protects eyes from radiation

tubes for food and water inside helmet

backpack with power supply and main oxygen supply

light, strong space suit keeps out cold and radiation

No Soil Toil ....

Perhaps the most important experiment to be conducted on a Mars mission will be trying to grow plants without soil. So, what do you grow the plants in, if not soil? The answer is a solution of water and nutrients. This technique is called hydroponics, and you can see how it's possible even without nutrients in the following experiment.

Move to Mars
GROW PLANTS WITHOUT SOIL

WHAT YOU'LL NEED

* some green or brown lentils
* a glass jar
* a pair of pantyhose
* an elastic band

WHAT TO DO

Fill the jar halfway with water, add two or three teaspoons of lentils, and leave it in a warm, dark place overnight. The next day, stretch a piece of pantyhose over the lid of the jar and fix it in place with the elastic band. Turn the jar upside down to strain out the water over a sink. Keep the jar in a warm, dark place and rinse the lentils with water a couple times a day.

WHAT HAPPENS?

After four or five days, some of the lentils should start to sprout both shoots and roots. When seeds are soaked in water and left in the dark, an alarm bell goes off inside the seed: "Time to grow!" The food supply inside the seed is enough to get the growing started. This is used up quickly, however, and when you travel to Mars, you will need to take nutrients with you to feed your plants as they grow.

You could also try to grow your plants in inflatable greenhouses. The soil can be imported from Earth, or it can be made from ground-up Mars rock with nutrients added. But where will the water come from? It will be transported from Earth, generated by fuel cells, or made from melted frost found under Mars's surface. Major investigations into the polar ice caps will have to be made to see if a major water source exists under the surface.

Polar Lander
Mars, 1999

Polar Lander reached Mars in December 1999, but scientists are still figuring out why it never returned their calls. This is a shame, because it was sent to study the soil and look for ice beneath the surface of the Martian south pole. The mission featured two microprobes, called *Deep Space 2*, designed to smash into the soil and look for signs of ice. The microprobes hit the ground at a speed of 447 mph (720 km/h). Ouch!

However water is first supplied, it will be fairly easy to produce more. Why? Because most of the water will be retained inside the Mars greenhouse. Want to see how? Try the following experiment.

Move to Mars
GROW PLANTS WITHOUT ADDING WATER

WHAT YOU'LL NEED
- ✳ a potted plant
- ✳ a clear plastic bag
- ✳ a twist tie

WHAT TO DO
Put the plastic bag over a leaf of the plant and gently tie it with the twist tie. Leave it for 24 hours and then take a look.

WHAT HAPPENS?
Inside the bag you will see water droplets. These have come from the leaf as it breathes out ("transpires" is the fancy word for it). Water travels through a plant from the roots to the leaves. Most of the water that is breathed out by the leaves can be recycled and fed back to the roots.

Water isn't the only thing that would be recycled and reused on Mars. Nothing can be wasted at the Mars colony, simply because it costs too much to get it there in the first place. Even the contents of a toilet can be treated with chemicals and then used as a sort of fertilizer for growing plants.

Power Play ....

Without power to drive the machines in a Mars colony, your mission will be over before it's even begun. Power will probably be supplied in two or more different ways. If one power provider stops working for some reason, another can kick in to compensate.

Where will the power come from? Well, there's a range of possibilities. You will probably bring a selection of fuel cells and batteries along from Earth. Before you even take off, experts will have told you whether it's going to be possible to make rocket fuel from the Martian atmosphere. If it is, then the landing craft

70

will come with an electricity generator that runs off that type of fuel. Another possibility is a small nuclear reactor.

The most readily available source of energy will be solar power. Mars is a lot farther away from the sun than Earth is, so the intensity of sunlight there is only half what we get on Earth. The thin atmosphere of Mars will help, however. It lets a lot more of the sun's rays through than Earth's heavy atmosphere. Solar energy will be an important power provider for people on Mars, even though dust storms will interfere.

Getting Around ....

Mars's rocky and unpredictable surface means that most of your travels will be on foot, close to the landing area. Mars colonists won't have a Porsche parked at the base, but you will probably have a rover.

The rover will have large inflatable wheels or tracks, as well as a special rocking body so that it can climb up steep slopes without toppling over.

The other likely Mars rover will carry scientific experiments. An unmanned sailplane will be launched by small rockets and kept in the air, possibly by solar energy. For longer journeys, this gliderlike craft will have its journey programmed in. Closer to base, it will be remote-controlled by astronauts.

Staying Long? ....

If your colony is a large one, and you plan to stay for a while, you will have to plan even more carefully. It will be impossible to transport everything a large colony needs from Earth. It will take a different approach.

The big idea in Mars studies at the moment is a philosophy called "living off the land." We know it will be impossible to live off Mars completely, but we might be able to use enough important materials from its atmosphere and ground to make life a lot easier for ourselves.

In the future, we might see oxygen and other useful gases being extracted from the Martian atmosphere, and possibly from water sources below the planet's surface.

There she blows! We've struck water!

In the long run, Mars's surface materials could be extracted and used. Martian soil contains large amounts of gypsum, which can be used to make glass. It might be possible to crush the soil to make bricks, clay, and cement. Silicon in the Martian soil could be turned into microprocessors—silicon chips!

CHAPTER SEVEN

TERRAFORMING

It's one thing to send small numbers of colonists to Mars to live in closed environments. It's quite another to make Mars a suitable place for millions of people to live. The technical term for shaping a planet to make it more like Earth is terraforming. No one knows whether terraforming is really possible. One day—perhaps not too far in the future—we'll be ready to try it. Results won't be seen overnight, though. Terraforming would take thousands of years.

Most people agree that there are three key things needed to get terraforming going:

* ☆ heat up the planet
* ☆ create a much thicker atmosphere
* ☆ introduce basic life forms

75

Brrrr! Turn Up the Heat......................

Heating up the planet would mean melting much of the ice believed to exist on Mars. It would also help thicken up the atmosphere. Several interesting ways of warming up Mars have been suggested.

(This illustration isn't one of them!)

The main idea is to trap more of the sun's energy on Mars. There are several different ways this could be done. Giant solar reflectors could be built and put into space. These reflectors, covered in a sort of high-tech aluminum foil, would redirect many of the sun's rays onto the polar ice cap, melting part of it and releasing gases into the atmosphere.

Another way would be to spray dark dust (possibly from Mars's moon, Phobos) onto the polar caps. Dark colors absorb more heat than light ones, so a thin layer of dust might be enough to melt the poles.

Move to Mars
TURN UP THE HEAT

You can see how aluminum foil reflects, and black materials absorb, the sun's energy with the following experiment.

WHAT YOU'LL NEED

- ✳ four jars
- ✳ a piece of black paper
- ✳ some aluminum foil
- ✳ a piece of white paper
- ✳ a few elastic bands
- ✳ a sunny day
- ✳ a thermometer

black paper   white paper   aluminum foil

WHAT TO DO

Wrap white paper around one jar, black paper around another, and aluminum foil around a third. Hold the materials in place with elastic bands, and fill the jars two-thirds full with water. Fill a fourth, clear jar with water and place all four in the sun. After an hour, record the water temperature in all four jars.

Creating the Right Atmosphere ....

Thickening Mars's atmosphere will be absolutely essential to successful terraforming. Heating up Mars will help release some gases into the atmosphere, but much more needs to be done. Many scientists think that a current problem on Earth might offer a solution to heating up Mars. Global warming on Earth is believed to be the result of the greenhouse effect. This is where carbon dioxide and water vapor in the atmosphere act like a blanket, trapping much of the sun's energy. This causes average temperatures to rise.

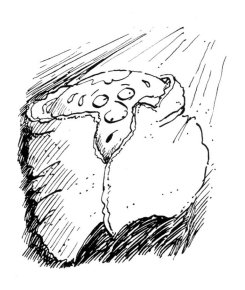

Earth's greenhouse effect is the result of hundreds of years of burning fossil fuels and cutting down large areas of forest. On Mars, it might be possible to create a similar effect by pumping large amounts of gases into the atmosphere. It would take a very long time, but the gases would thicken the atmosphere, enabling it to trap more heat and warm the planet a little more. A thicker atmosphere would mean more atmospheric pressure—something that most plants need to grow.

This all sounds great, but where are these gases going to come from? Well, they can't all be transported from Earth. Perhaps processes will be invented to make the gases using materials from Mars's surface. Melting parts of the polar ice caps, which contain carbon dioxide, will help.

One extreme-sounding idea for adding gases to the atmosphere involves using comets, which are made of solid ice and hang out in the outer parts of the solar system. The future might see several comets redirected so that they plunge into Mars's atmosphere and burn up, instantly adding gases.

Hope for Japan

Mars, 2003

Mars's atmosphere, the little there is of it, will shortly get a thorough checking from the first Japanese mission to Mars. The *Nozomi* (which means "Hope") spacecraft is intended to stay at the red planet for a whole Mars year, to measure and map the atmosphere, assess Mars's weak magnetic field, and beam back photos of dust storms and the polar ice caps.

Engine problems, which prevented the craft from arriving in 1999, have been fixed. *Nozomi* should reach Mars in 2003.

Introducing Basic Life Forms...............

We don't mean introducing basic life forms to each other, but rather introducing them on Mars. Perhaps "basic" is unkind: these plants and microscopic creatures are among the hardiest of all the things that live on Earth. That's why they would come first—but only after Mars has warmed up a little and started to build its atmosphere. Earth soil is a good source of bacteria and microbes, as the next experiment shows.

Move to Mars
SEE SOIL BREATHE

WHAT YOU'LL NEED
* ✳ some garden soil
* ✳ a small dish of limewater
  (ask at school or
  pharmacy)
* ✳ a plastic box with a lid

WHAT TO DO
Place a few tablespoons of soil
in the box and then put a
small dish of limewater inside, next to the soil. Put the lid on
the box and leave.

WHAT HAPPENS?
Limewater is a liquid that gets cloudy if there is carbon dioxide
gas present. After a couple of days, the limewater will turn
cloudy, showing that there is carbon dioxide in the box. It
comes from thousands of tiny creatures in the soil that breathe
out carbon dioxide.

Some of these microbes, and certain plants, can
survive in low temperatures with little moisture and
low atmospheric pressure. The bacteria and hardy
plants that make the first trips to Mars will be their
relatives, but they will be genetically modified to be
even more comfortable in Martian conditions and to
produce as much oxygen as possible. Dark-colored
plants will become a specialty because as they flourish
and cover parts of Mars, they will absorb more rays
from the sun and help heat the planet further.

Over time—and we mean thousands of years—larger plant and animal species may be introduced to Mars. The planet might even produce its own unique plant and animal life. We won't see it, but future generations might.

CHAPTER EIGHT

LET'S GET COLONIZING

You've handed in the last of your homework and canceled the newspaper. You're ready to travel with the other members of the first Mars colony. How will it all happen?

Step 1: On Target ....

Well, first of all, there will be lots of missions by the types of probes and landers you've already read about. These will map the surface of Mars in incredible detail, which will help scientists choose the site of the first colony. More missions, equipped with advanced robots, will perform lots of experiments and try out new equipment. Testing this gear is very important.

Prototypes of rocket-fuel makers and plant-growing kits must work well for a manned Mars colony to be realistic.

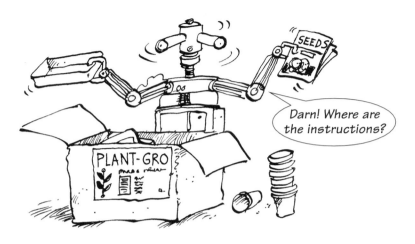

The possibility of making rocket fuel on Mars will be tested out very soon. One of the experiment packages onboard the *Mars Surveyor*, which will land on Mars in 2002, will see if it is possible to synthesize rocket fuel from the gases in the Martian atmosphere.

Mars Surveyor is a step up from the *Pathfinder/Sojourner* mission. It too will feature a robot crawler, deployed by a robot arm, which will explore part of the Martian surface. The whole mission is designed to measure the dangers human explorers would face on Mars.

Step 2: The Go-Ahead ....

After much research, and perhaps even more missions to test it all out, the Mars colony will get the big thumbs-up. Plans for the human missions will really start rolling.

First, large fleets of unmanned spacecraft carrying robots and essential equipment will head off to Mars. Landing at the target site, the robots will build and test the gear needed for the first Mars colony. Some will be mobile rovers, which will scan every square inch of the target site, mapping the features in minute detail.

The robots will also assemble buildings and equipment. Supervisor robots will relay pictures and other information back to mission control on Earth.

Step 3: Train Your Brain ....

Back on Earth, the excitement will be building. As one of the first potential colonists, you will be an instant celebrity.

You won't have much time for talk shows or newspaper interviews, though.

You'll be much too busy training. You'll need to be physically fit, to understand all the procedures of the mission, and to know your own responsibilities inside and out. Each member of the first colony team will be a mission specialist with his or her own important job. You could be the in-flight navigator, the crew's doctor, a space engineer, or a geologist. You'll also have to learn the key parts of another crew member's job, so that if one person gets sick, there's always a replacement.

I know you're the ship's doctor, but I said to operate the engines, not operate ON the engine.

Step 4: Liftoff! ....

The launch window has arrived. Finally, the green light will come. You'll be allowed to say goodbye to your nearest and dearest, then it's time to get into the craft, buckle your seat belt, and take a deep breath. You're off to Mars!

Did I pack my toothbrush?

Step 5: In Flight................................

Liftoff is the most exciting ride of your life. You feel the forces thrusting you back into your seat as the giant rockets propel you out of Earth's orbit. After the euphoria of a successful liftoff, the next few hours bristle with tension. Is everything working okay? Is the mission on course? Has mission control detected a problem? Is the craft going to be called back to Earth?

Once you know all is well and the ship is on course, life settles down. You are on a journey that will make a four-hour trip in the school bus seem like a picnic. Six months or more stand between you and Mars. Hope you packed a few good books on tape!

Actually, you've got plenty to do. You're keeping fit by working out in space, you have assignments to perform on the ship, and you're keeping yourself in training for your vital tasks on Mars…

…but not necessarily all at the same time.

Mission Control is like an annoying parent. They want to know absolutely everything about you day in and day out. This includes physical stuff, such as blood pressure, but also how you're feeling.

Step 6: Happy Landings ....

Gradually, the excitement aboard the ship builds as
you get closer and closer to your destination. Entering
Mars's orbit and landing on the surface is the most
dangerous part of the mission. Powerful computers
handle most of the descent stage, but the crew has an
override button in case something goes wrong. If all
goes well, a combination of parachutes and brake
thrusters will land you softly and safely on Mars.

Step 7: Those First Days on Mars ....

You won't be heading out onto Mars immediately.
There will be a day or so of checks and double-checks
before you slip into your special space suit. Which of
you is going to be first down the ladder and onto the
surface will have been decided months ago.

Whether you're first, second, or eleventh, you'll never forget your first walk on the surface. But even that first walk will involve jobs such as checking the outside of your landing craft. Over the next few days, you and the rest of the crew will venture farther afield, perhaps in your very own Martian rover.

Finding and checking the equipment set up by the robot missions will be top priority. Then you and the rest of the crew will start assembling more experiments and equipment, such as inflatable greenhouses for growing plants.

Step 8: Settling Down ....

You will spend several months on Mars. Then again, if you are very lucky—all the equipment works, water has been found, energy is easy to create, and plants can be grown—you'll be in for a longer stay. There's even a chance you'll be joined by more Mars colonists.

By the time more people arrive, you'll be living in large buildings connected to the outside by airlocks. Every trip outside means several hours of getting into your space suit, getting used to the pressure in the airlock, and cleaning yourself and the airlock afterward.

But it won't be all hard work. However badly mission control back on Earth wants to get value for its money, they know they can't work the first colonists into the ground. Leisure time will be carefully designed to keep you fit and happy.

Step 9: Terraforming ....

Terraforming will start after several centuries of Mars colonies. Experiments and fieldwork will reveal the best way to thicken the atmosphere, release water supplies, and encourage natural plant life. Work on a scale never attempted before will gradually transform the red planet into a blue and green one. Work on terraforming might get a boost from advances in nanotechnology. This is the science of making machines on an incredibly tiny scale. How small?

Well, they're measured in atoms rather than inches. Millions of mass-produced nanobots would become a willing workforce capable of transforming Mars into a lush world.

Step 10: The New World....................

Many thousands of years after you arrive, Mars is unrecognizable as the cold, hard, rocky world you knew. The terraforming process has worked its magic, turning Mars into a planet with flowing water, a thick atmosphere, and lush plant life. Dozens of generations of Mars-dwellers have come and gone. The unique Martian diet, low gravity, and other factors have produced variations in the way people from Mars and Earth evolve. Who knows—by that time, the human race might be in cahoots with alien species!

The thousands of Mars settlers know all about you and the rest of the first colonists to arrive. You and your crewmates from that first-ever Mars mission have been elevated to the status of gods.

Then again, perhaps your name is forgotten, just like the very first caveman or woman who looked up and spotted a strange red star twinkling in the night sky.